AMAZING BODY SCIENCE

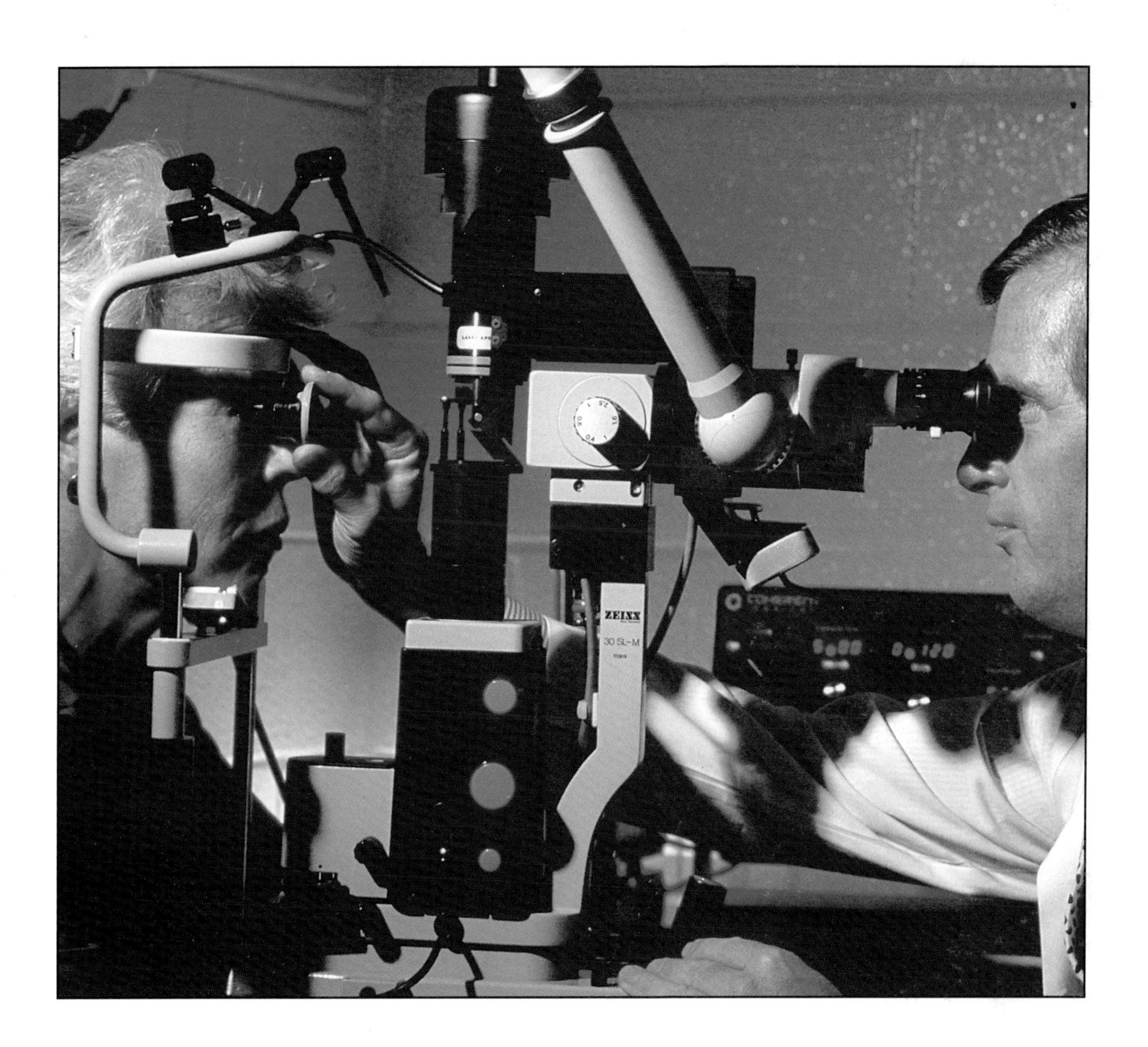

Edited by Nicole Carmichael

Written by Rosalind Lowe

WORLD BOOK / TWO-CAN

AMAZING BODY SCIENCE

First published in the United States in 1996
by World Book Inc.
525 W. Monroe
20th Floor
Chicago
IL USA 60661
in association with Two-Can Publishing Ltd.

**For information on other World Book products,
call 1-800-255-1750, x 2238.**

ISBN 0-7166-1737-4 (pbk.)
LC: 96-60467

Printed in Hong Kong

1 2 3 4 5 6 7 8 9 10 99 98 97 96

Design by Simon Relph. Medical consultant: Dr. Mike R. Johnson, D.Phil, MRCP (Lon). Picture research by Debbie Dorman. Production by Joya Bart-Plange.

Front cover photographs: Science Photo Library. Illustration: Annabel Milne.

Picture credits: Two-Can: 4, 9tl, 14tc, 22/23, 29tc, 31. Science Photo Library: 6/7t, 18t, 19, 25, 28tr&br, 29tl,bl&br. Rex Features: 12. Metropolitan Police: 14b. Kobal Collection: 24. P.A News: 26-27tc. South West News: 26-27br.

Illustrations: Annabel Milne: 5, 6/7, 8/9, 15, 16/17, 20/21. Woody: 7, 27. Simon Relph: 7, 18, 25. Bernard Long: 10/11. Chris West: 13. Russ Carvell: 30.

CONTENTS

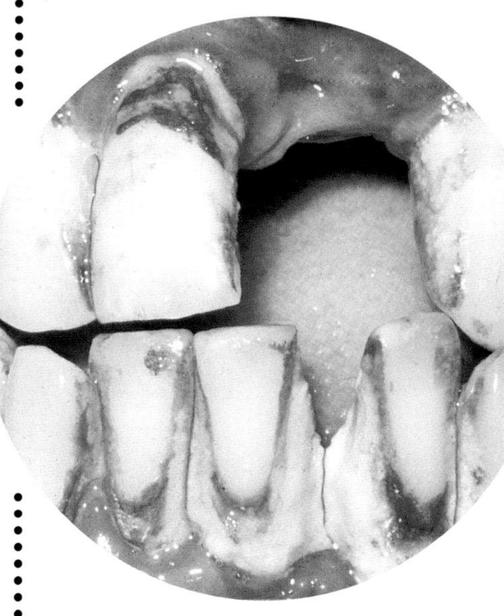

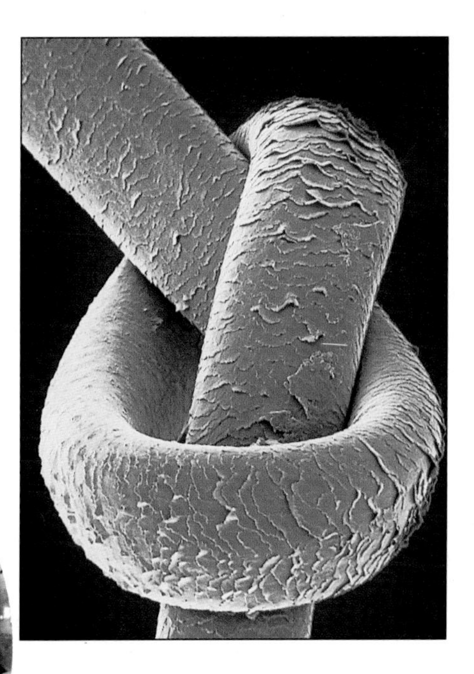

HAVE A GUESS...

A) WHICH OF YOUR FINGERNAILS GROWS THE FASTEST?
B) WHERE IS THE HOTTEST PART OF YOUR BODY?
C) HOW MANY MUSCLES DO YOU USE WHEN YOU RAISE YOUR EYEBROW?

YOUR TIME'S UP. THE ANSWERS TO THE ABOVE QUESTIONS ARE:

A) YOUR MIDDLE FINGERNAIL
B) THE CENTER OF YOUR BRAIN
C) ABOUT 30 MUSCLES

Your body is a mega machine, capable of amazing physical feats. It's more sophisticated than any human-made machine on the planet. It runs smoothly for years and can even repair itself when things go wrong.

But how much do you know about how your body works? Is your knowledge skin deep or are you a body whiz? Test yourself with the following teasers, then bone up on bones, blood, and all your body's other working wonders.

1 Which part of your body is the least sensitive to pressure?

2 Which are the coolest parts of your body?

3 Which is your longest bone?

4 Which is your heaviest organ?

5 How far would your blood vessels stretch if you placed them end-to-end?

6 From which part of your body does the most heat escape?

7 Which human bone is broken the most often?

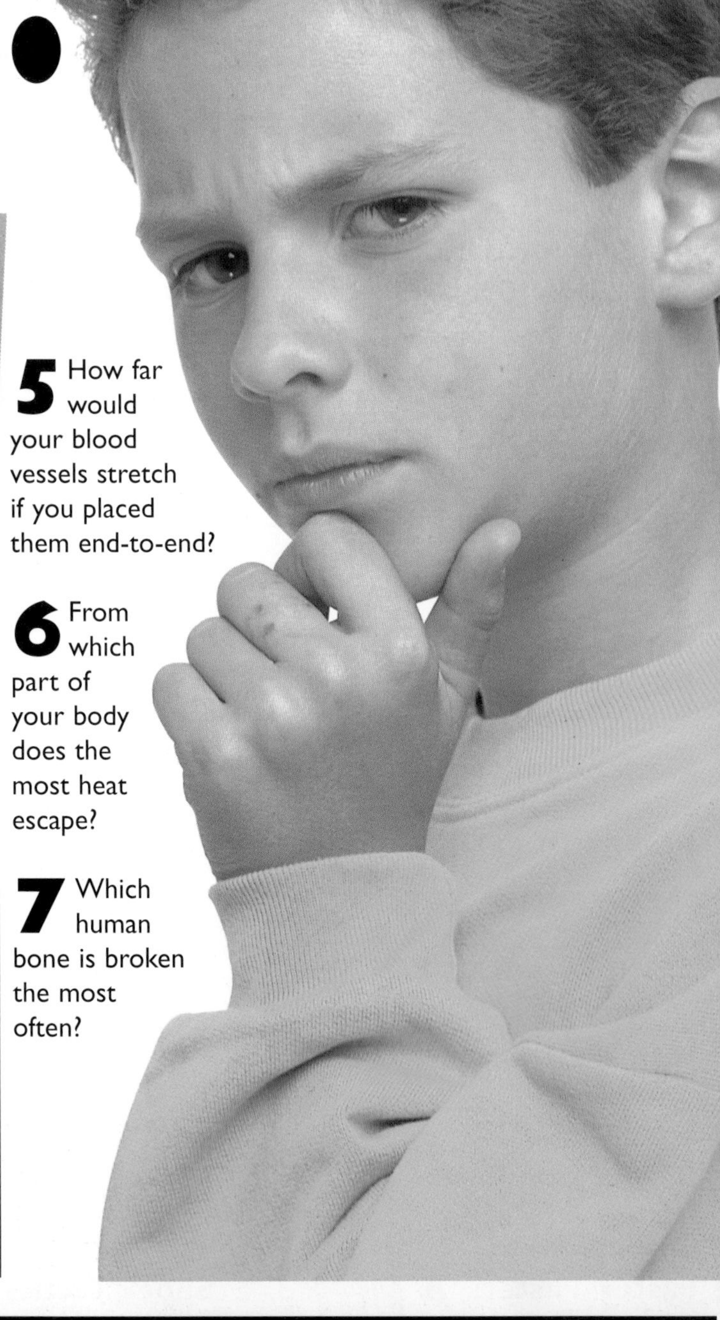

ANSWERS

1. Your bottom – which is a good thing, as you spend a lot of time sitting on it!
2. Your fingers and feet
3. Your thigh bone
4. Your skin – it weighs 8.8lbs
5. Around 60,000 mi – that's twice the circumference of the earth!
6. Your head
7. Your collarbone

The average human is made up of all this!

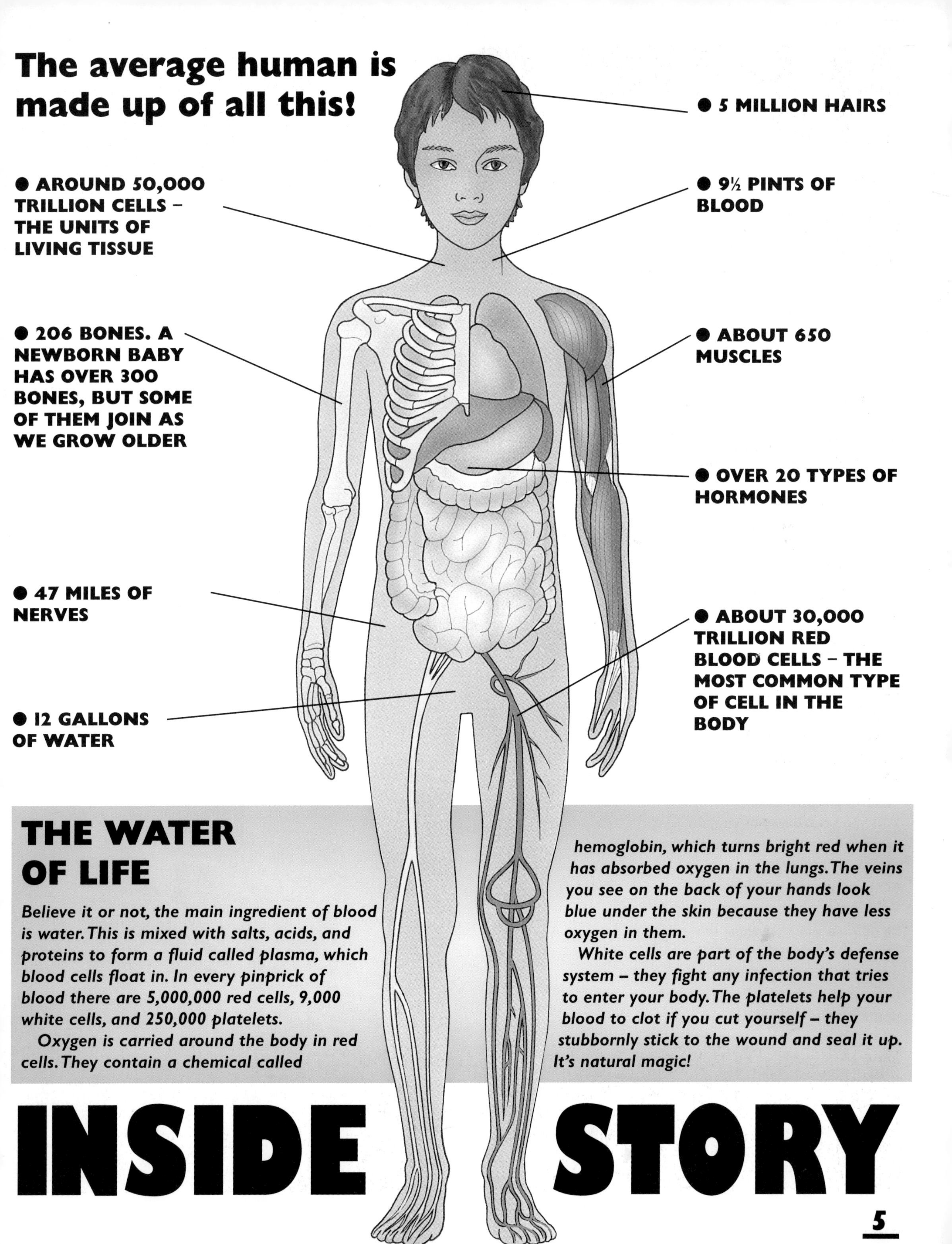
5 MILLION HAIRS
AROUND 50,000 TRILLION CELLS – THE UNITS OF LIVING TISSUE
9½ PINTS OF BLOOD
206 BONES. A NEWBORN BABY HAS OVER 300 BONES, BUT SOME OF THEM JOIN AS WE GROW OLDER
ABOUT 650 MUSCLES
OVER 20 TYPES OF HORMONES
47 MILES OF NERVES
ABOUT 30,000 TRILLION RED BLOOD CELLS – THE MOST COMMON TYPE OF CELL IN THE BODY
12 GALLONS OF WATER

THE WATER OF LIFE

Believe it or not, the main ingredient of blood is water. This is mixed with salts, acids, and proteins to form a fluid called plasma, which blood cells float in. In every pinprick of blood there are 5,000,000 red cells, 9,000 white cells, and 250,000 platelets.

Oxygen is carried around the body in red cells. They contain a chemical called hemoglobin, which turns bright red when it has absorbed oxygen in the lungs. The veins you see on the back of your hands look blue under the skin because they have less oxygen in them.

White cells are part of the body's defense system – they fight any infection that tries to enter your body. The platelets help your blood to clot if you cut yourself – they stubbornly stick to the wound and seal it up. It's natural magic!

INSIDE STORY

MASTER MIND

Your brain might look like an over-grown walnut, but it's 100 percent pure genius!

Your brain is your body's central control-center. As well as thinking and remembering, it directs all of your body's movements and feelings.

And even though it's only a tiny fraction of your total weight, it's the most powerful organ in your body.

Over three-quarters of the brain is made of water!

TOP SIX

You may not reach your full height until you're well into your teens, but your brain reached its full weight of about 3 pounds by the time you were just six years old. During that time you learned new skills, such as talking and walking, at the fastest rate of your life – it's no wonder you needed so much brain power!

When you're awake your brain is constantly receiving masses of information, some of which is stored as memories. Without memory you wouldn't be able to learn anything new, and you would have the same experiences for the "first time" over and over again. Every time you met a friend, you'd have to start getting to know him or her from scratch.

GRAY MATTERS

People often call the brain "gray matter," but it's not just a nickname. The outer part is actually a grayish pink color

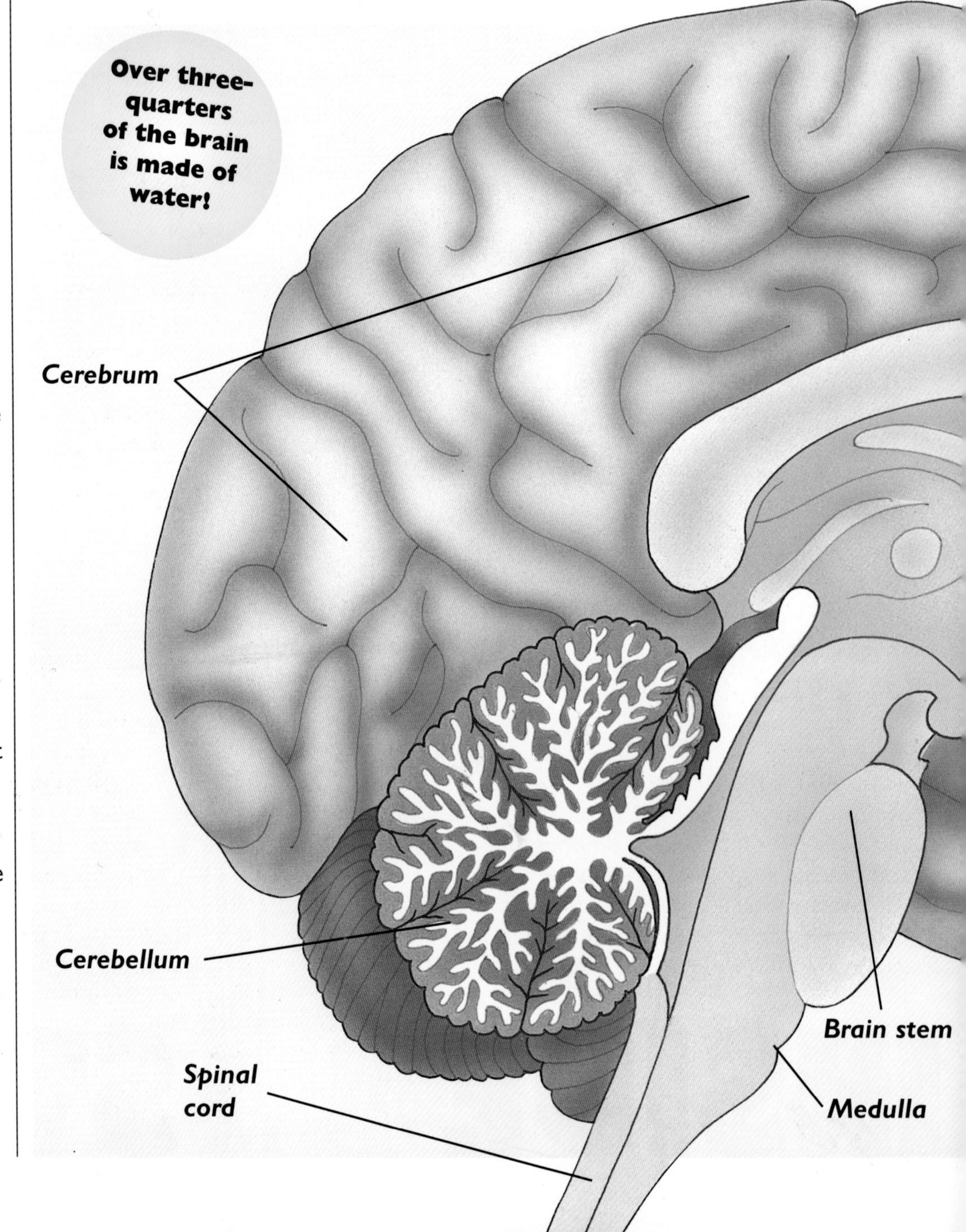

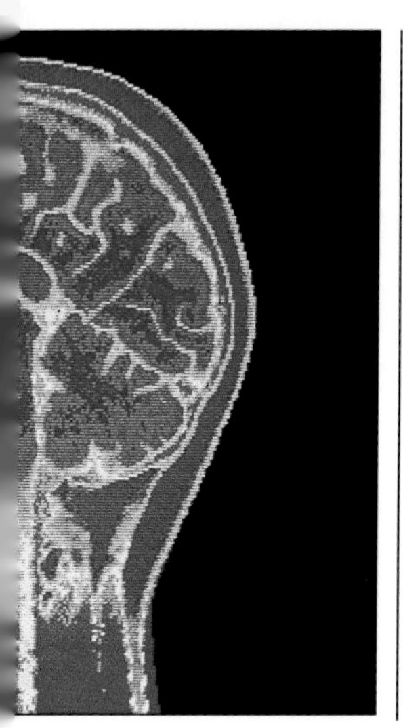

and consists of millions of nerve cells. The white inner part is made of nerve fibers that connect with other parts of the nervous system.

Your gray matter (or cerebrum) receives important messages from the body's sense organs and sends out nerve impulses that control all the body's muscles. It also analyzes and stores information. In fact, you're using this part of your brain when you read this page!

Because of its folded, wrinkly shape, a lot of brain can fit into a relatively little skull

The cerebellum is found underneath the cerebrum, at the back of the head, and is responsible for balance and co-ordination. When you learn to play an instrument or ride a bicycle, this part of the brain is learning to control the many muscles involved, so you can soon do these things "without thinking." If you "think" about doing them, you may slip up when you became aware of how complicated these activities are – which is why your body tends to let the cerebellum get on with things!

The brain stem is where the brain merges with the nerves of the spinal cord, which carries messages to the rest of the body. Below this is the most vital part of all. The medulla controls reflexes like sneezing as well as regulating important functions, such as your heart rate, blood pressure, and breathing, even when you're asleep. In fact, you don't even have to think about using your brain: it thinks for itself.

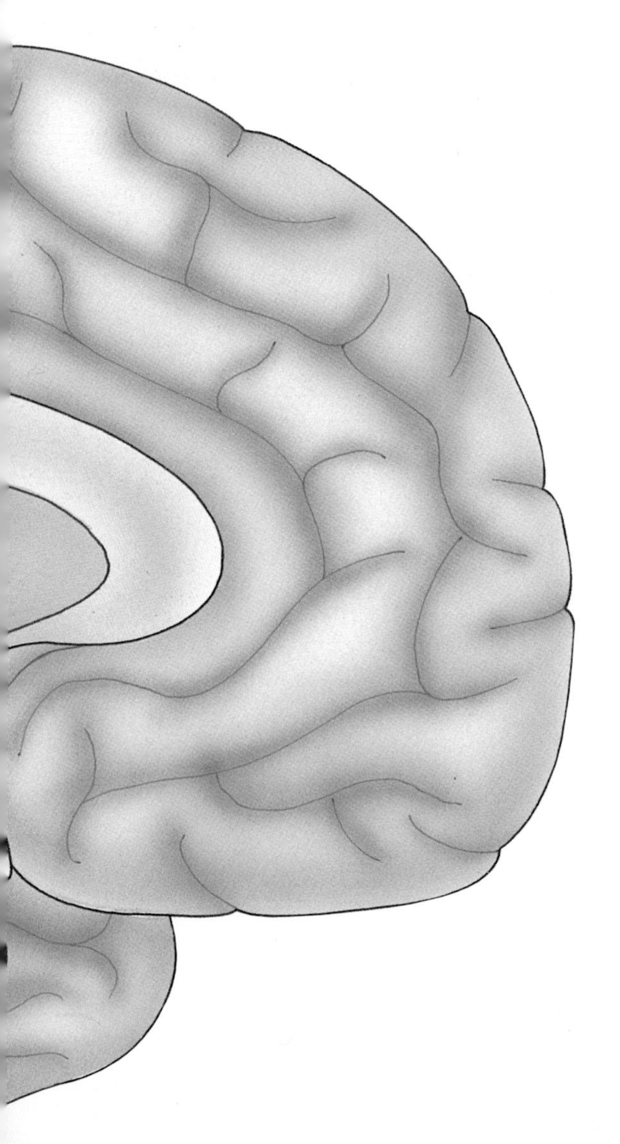

RIGHT-HANDED OR LEFT-HANDED?

If you are right-handed, you use the left-hand side of your brain to do logical things, such as reading and solving math problems. But if you're left-handed (one in ten people are), then it's the right-hand side of the brain that does the logical thinking.

Meanwhile, your creative and imaginative side is controlled by the other half of your brain. That's the right-hand side for right-handed people and the left for left-handed people.

AMAZING MEMORIES

Have you got a good memory? Maybe you're good at remembering historical events or athletes' names. But could you memorize the order of a pack of cards? British mega-brain Dominic O'Brien can. In 1993 he remembered the order of not one, but forty packs of cards that had been shuffled together. After a single sighting, he made just one mistake when reciting the order of the 2,080 cards.

SENSE AND SENSE

Sight, hearing, touch, taste, and smell — the five vital senses that make sense of the world around us

From the moment you open your eyes in the morning, around a million pieces of information are sent to your brain every second about what you see around you. In fact, the human eye is so sensitive that experts believe it's possible to detect a single burning candle up to 50 miles away. What a sight for sore eyes!

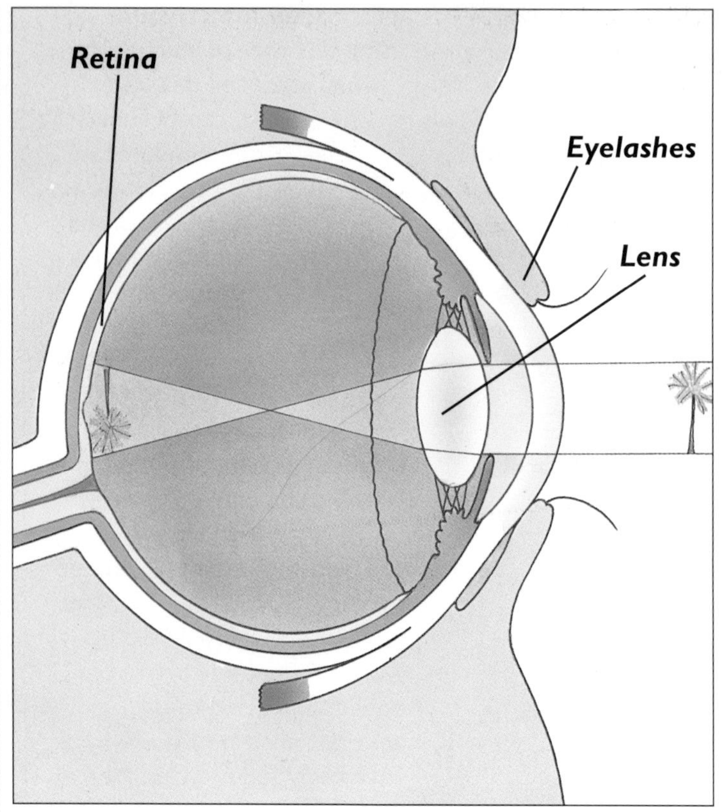

SEEING IS BELIEVING

The human eye can see around 10 million different colors and shades. But in actual fact, your eyes don't really "see" objects. They see the light that objects give off. That explains why you can't see in the dark.

Light enters your eye and passes through a transparent lens that bends the rays into a point on the retina (the back wall of the eye). These sensitive cells transmit impulses to the brain. The image of what you see is formed upside down on the retina, but your brain cleverly turns it right side up.

BROWN, BLUE, OR GREEN EYES?

Eye color is determined by the amount of melanin pigment in your body. Large amounts produce the brown or hazel color, and smaller amounts result in blue or light green eyes.

SOUNDS SURPRISING!

Your ears hear sounds by picking up vibrations of air. For instance, when a bell rings it causes the air around it to vibrate, and these sound waves are funneled to your eardrum via your ear. Linked bones in your middle ear pick up the eardrum vibrations and transmit them to a coiled tube called the cochlea. Here the vibrations are picked up by tiny hairs and turned into nerve messages that travel to the brain.

HOW LOUD IS LOUD?

The unit used to measure the intensity of sound is the decibel. Very quiet sounds like whispering have low decibel numbers, but sounds of 140 decibels or more cause pain in the ears and may even damage them.

ACTION	DECIBELS
Whispering	20
Normal talking	60
Vacuum cleaner	80
Loud shout	90
Road drill	115
Amplified band	120
Jet taking off	140

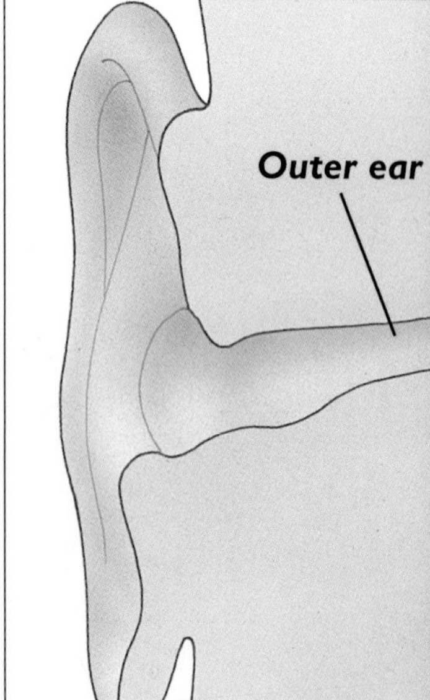

NOT TO BE SNIFFED AT

Amazingly, your nose can detect over 4,000 smells, and the sense of smell is estimated to be about 10,000 times more sensitive than the sense of taste.

When you breathe in through your nostrils, the air passes over a patch of nerve endings situated high in your nose. Any odors present dissolve in the sticky mucus there. Then hairlike sensors set off a chain of impulses that travel up the nerve fibers to the brain. There they are translated into different smells. Appetizing odors can make your mouth water, but no one's quite sure how we tell good and bad smells apart. Some scientists believe that there are several types of nerve endings for smell. These respond to a particular type of odor and relay messages to your brain telling it if it's a good or bad smell.

Cochlea
Middle ear
Inner ear

SMELLY MEMORIES

Does a particular smell remind you strongly of a special person or day? Smell can bring back old memories because it is recognized in the part of your brain that also deals with memories and feelings.

A QUESTION OF TASTE

By detecting tastes, your tongue makes eating more enjoyable and can also stop you from swallowing something unpleasant. Its undersurface is smooth, but many small lumps (papillae) give it a rough surface. On or between these papillae are around 9,000 taste buds. Their nerve endings react to particles dissolved in saliva and send messages to your brain.

You may be able to tell many different tastes apart, but each taste bud can only detect one of four basic tastes. Your sweet taste buds are on the tip of your tongue. Salty tastes are picked up by the sides of the tip of the tongue. Meanwhile, the rear sides of the tongue detect sour tastes, and the back of the tongue picks up any bitter tastes.

Bitter
Sour
Sweet
Salty

TASTY SMELLS

Your senses of smell and taste are very closely linked. When you're eating a meal, it's your more sensitive sense of smell that picks up the subtle flavors of the dishes. Have you noticed how plain most food tastes when you have a cold and your nose is stuffed up?

TOUCHY SUBJECT

Your skin can detect pain, pressure, heat, and cold. That's because it contains millions of sensitive nerve cells that react to different sensations. Some touch receptors are found near the base of hairs, so that they can sense any slight movement in the shaft – that's why you might feel a minute insect crawling on your arm or the wind on your cheeks.

Your sense of touch is more sensitive in some parts of the body than in others. It's most highly developed on the tip of the tongue and least sensitive on the back of the shoulders. Your skin also feels colder in different places, such as on your fingers and toes, where there is less fatty insulation to keep you warm.

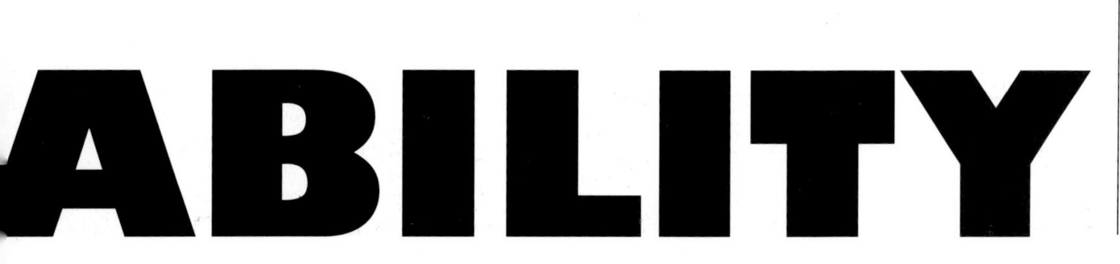

THE ROAD TO RECOVERY

Medicine's early pioneers

◀ DISEASE KILLER

In 1796, British doctor Edward Jenner developed the world's first vaccine – it was a way of stopping people from catching a terrifying disease called smallpox.

Jenner heard that dairymaids who had caught a less serious disease called cowpox didn't go on to catch smallpox. He decided that there must be a link between the diseases, so he came up with a clever experiment. By putting some infected cowpox matter on the arms of a young boy, then repeating the process a few weeks later with smallpox, Jenner discovered that the boy didn't develop the smallpox disease. The boy had been made immune to smallpox through his contact with cowpox.

Since Edward Jenner's first vaccine, smallpox has been totally wiped out, and now injections are used to prevent many other serious diseases, such as tetanus and tuberculosis.

Today many illnesses can be cured and some can be prevented altogether. But vaccinations and other modern medical miracles are only possible because of the dedication and discoveries of early medical researchers. Since then medicine has made significant advances.

▶ BLOOD LINE

In 1616, an English doctor, William Harvey, discovered how blood circulates around the body. By dissecting dead humans and animals, Harvey worked out that the heart produces a continuous stream of blood that circulates around the body, then returns to the heart.

Harvey's discoveries acted as a map of how the body works, a blueprint for further discoveries.

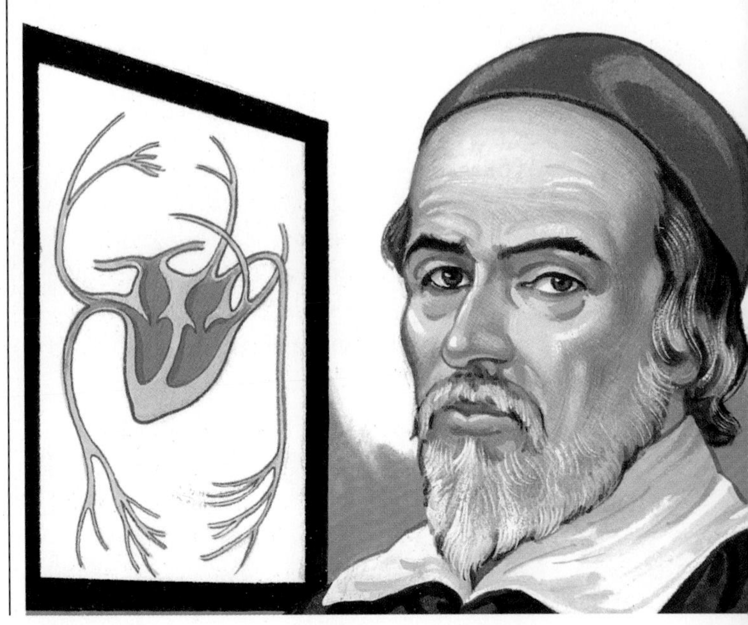

◀ LEADING LIGHT

Nurse Florence Nightingale was the founder of modern nursing. Although she came from a wealthy family, she set her heart on helping other people by studying nursing. When Great Britain and France went to war with Russia in the Crimea in 1854, Florence took charge of nursing sick and injured soldiers.

Conditions were terrible but Florence organized a cooking and cleaning system and demanded better supplies. At night, she patrolled the hospital corridors and was nicknamed the "Lady with the lamp." After the Crimean War, Florence became a famous expert on how to care for the sick.

▼ X-CELLENT!

In 1895, Wilhelm C. Roentgen, a German scientist, realized that if he passed electricity through a black tube it gave off an eerie green glow. He then discovered that these rays affected photographic film. He didn't know what to call these special rays, so he labeled them X, the scientific name for the unknown. Unlike light, the tiny particles that made up the eerie glow actually traveled right through living tissue.

When Roentgen perfected his photographic technique in 1907, doctors were able to use these X rays to look inside sick people and get a much clearer picture of their medical problems. Clearly a brilliant idea!

▲ CLEAN CUT IDEA

English surgeon Joseph Lister realized that germs were responsible for many patients' deaths during surgery. In 1865, he began to use a powerful disinfectant called carbolic acid to sterilize surgical wounds. Later, he insisted that surgeons wash thoroughly before an operation and wear gowns and masks, to stop germs from infecting wounds.

BREAKING THE MOLD ▶

In 1928, Scottish scientist Sir Alexander Fleming discovered how bacteria was killed by mold growing in his laboratory. This germ-blasting power was later developed by scientists Howard Florey and Ernst B. Chain to make the life-saving antibiotic penicillin, which is now used to fight pneumonia and other infections caused by bacteria.

For the brilliant germ of an idea, Fleming, Florey, and Chain were awarded a share in the 1945 Nobel Prize for medicine.

HOW FAST DOES

Olympic athletes may be the fastest things on two legs, but that's nothing compared to the speeds achieved by other parts of the human body. Messages travel to your brain at approximately 180 miles per hour, your heart beats 150 times a minute when you're exercising, and you can sneeze at a mega-speedy 104 miles per hour!

Even growing is pretty fast. At birth the average baby boy is around 20 inches long. By his second birthday he will have sprouted another 14 inches, and then he will continue to shoot up by about 2 inches a year until his teens. Then boys go through another rapid growth spurt between the ages of 14 and 15, and carry on creeping up the height chart until they are about 18. Girls might not be as tall in the long run, but they do their growing at an even faster rate.

Meanwhile, the other parts of your body are ticking along at a steady pace. Your blood is being pumped from your heart at an even 2.2 miles per hour, while you talk at around 11 miles per hour, and laugh at 15 miles per hour. But when your reflexes spring into action, they really move!

Try beating these records: 100 finger-tapping taps in 15 seconds and 70 super-quick blinks in 15 seconds. Phew!

OUCH!

Have you noticed that if you prick your finger, your body moves your hand away from this danger before your mind's even had time to register it? That's because the pain impulse travels mega fast along your nerves – about 250 miles per hour faster

YOUR BODY GO?

than a high speed train! When it reaches your spinal cord, a speedy instruction is immediately relayed to your arm, telling it to move double-quick! You instantly move your finger out of danger. This is called a reflex action, because you do it without thinking. Unfortunately, the pain message still travels to your brain even after your finger is safe; that's why it hurts a second or so later.

THE RACE IS ON!

If you reckon you're quick off the starting block, check out these awesome animals!

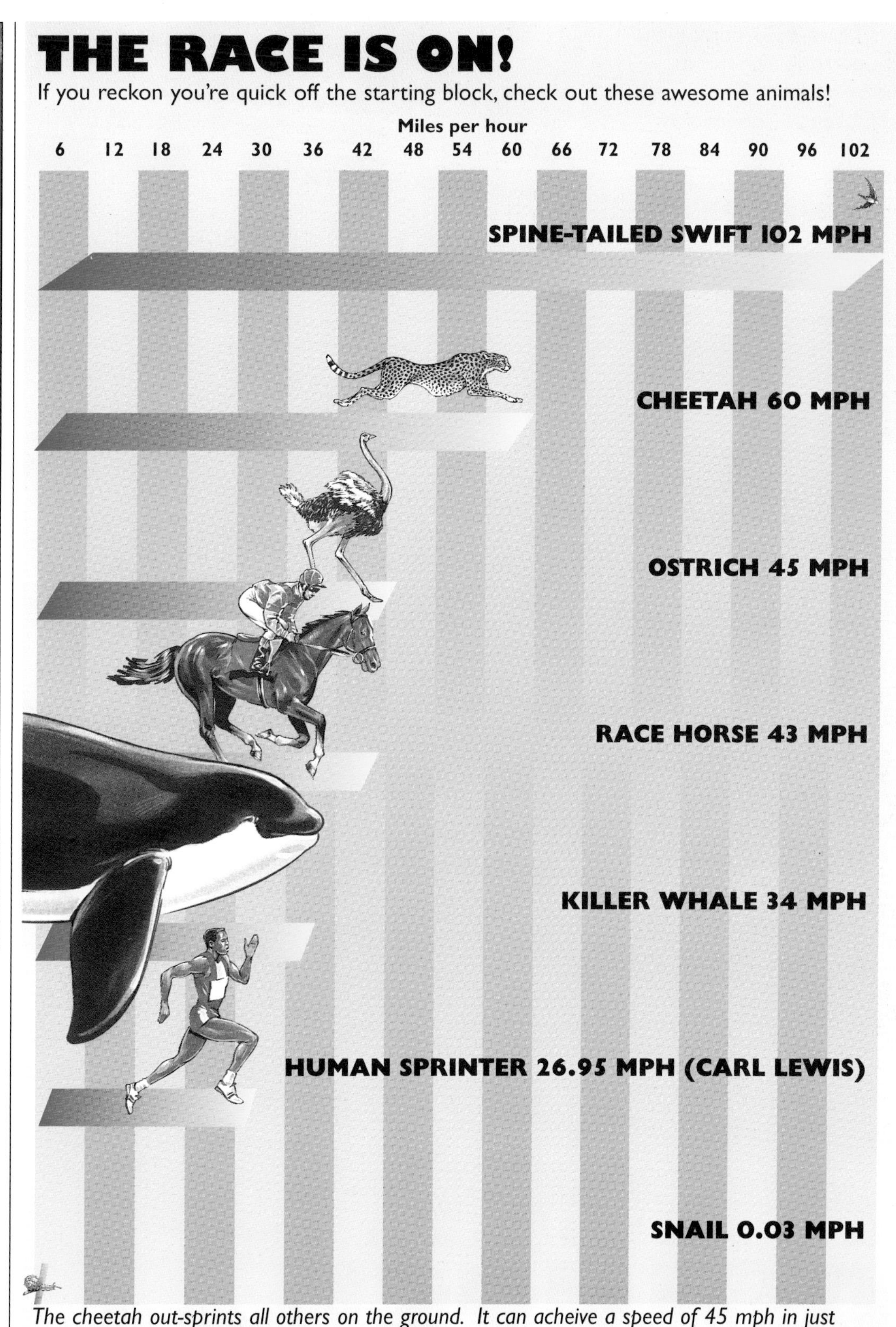

The cheetah out-sprints all others on the ground. It can acheive a speed of 45 mph in just TWO seconds. In the air the spine-tailed swift is amongst the high-flyers.

THE FINGER POINTS!

Focus on fingerprints

Look closely at your fingertips and you'll see lots of tiny ridges on your skin. The patterns they make are totally unique to you – even if you have an identical twin, that person's fingerprints will be different from yours!

This special pattern was formed before you were born and will stay the same throughout your entire life.

▲ ***Fingerprints have to match in at least 16 ways before they can be used as evidence of a crime.***

TAKING SHAPE

Even though everyone's fingerprints are different, they tend to fall into three general patterns.

The most common pattern is the curved LOOP. It has curved ridges that begin and end on one side of the finger. The second most common is the swirly, circular WHORL. And a few people have a plain ARCH, where ridges rise ever-so-slightly in the center. But the chances are that the patterns on your fingerprints are a combination of all three.

Arches

Loops

Whorls

Our fingerprints date back to our tree-living ancestors, who had friction pads on their fingers and toes. These folded areas of skin helped them to grip as they swung around the forest.

All we have left of the friction pads now are the small ridges of our fingerprints. As you touch something a minute trace of sweat is left on the surface in your own fingerprint pattern.

LOOKING FOR CLUES

In the 1890s, a sharp-eyed British police officer named Edward Henry invented a new crime-cracking method based on these fingerprint patterns. By studying the exact fingerprint shapes, Edward Henry discovered that criminals made their mark at the scene of a crime. He then set up the first Fingerprint Branch in Britain and, in just six months, over 100 successful identifications were made.

Since then, fingerprinting has become a vital element in detective work. If prints aren't clearly visible at the scene of the crime, police dust the area with aluminum powder or add chemicals to show up any fingerprints. Then it's a bit like doing the hardest spot-the-difference puzzle in the world. And before the prints can be used in evidence in Britain, they have to match each other in 16 different ways.

Criminals try to beat the system by having drastic plastic surgery to remove their fingerprints, but everyone has distinctive markings on other parts of their fingers, too. Even wearing gloves doesn't baffle detectives – the super sleuths have developed a way of tracing glove prints. Very handy!

GENE GENIUS

Today's sleuths are using another unique body pattern to track down suspects. This is DNA – your genetic recipe, or individual body "code." DNA is found in every single one of your cells and carries all sorts of vital information about your body.

By testing samples of, say, hair or saliva, DNA experts can find out if this pattern matches clues found at the scene of a crime.

In 1995, the world's first computerized DNA database was set up to keep records of suspects' DNA blueprints. Detectives hope that by the year 2000, five million DNA records will be kept there.

FOOD FOR THOUGHT

When you eat a meal you're sending it on a journey that will last between 24 and 36 hours. During that time, it will travel along 30 feet of pipes in your body and be broken down into minute particles. This is called digestion, and it's a very impressive process!

START HERE

The meal's long journey begins in your mouth. Here your front teeth bite off chunks of food before your back teeth grind them into a pulp that can easily be swallowed.

When you smell, taste, or even just think about food, your salivary glands flow into action, pouring saliva into your mouth. This contains special substances called enzymes, which start to break down the food while it's moved around by your tongue and jaw muscles.

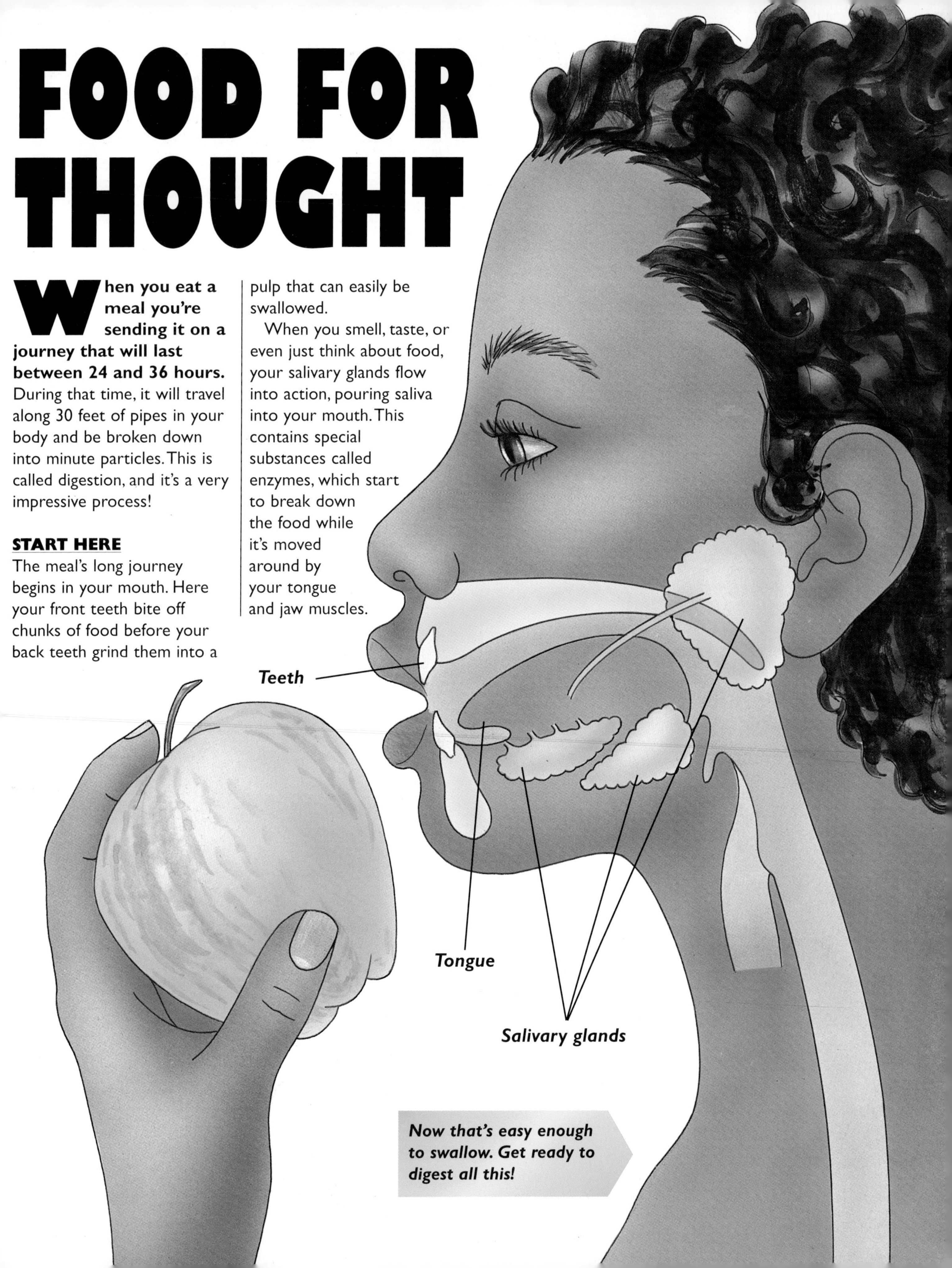
Teeth
Tongue
Salivary glands
Now that's easy enough to swallow. Get ready to digest all this!

THE JOURNEY CONTINUES...

The digestive system in action

When food is swallowed, the tongue pushes it down into an 8-inch-long tube called the esophagus. The muscles in the walls of the esophagus contract to push food toward your stomach – this means you could swallow food while standing on your head if you wanted to!

Your stomach is a flexible, J-shaped bag that can hold up to 2 quarts of food. Some foods stay here for five hours, where powerful muscles in the stomach walls churn the contents about and mix them with gastric juice.

The gastric juice contains enzymes that break down proteins, plus an acid that kills off any germs you may have swallowed. If your stomach is empty, it fills up with gas, and those churning movements make your stomach rumble.

The food and gastric juice mixture is known as chyme. It is squirted into the small intestine, which at 16 feet long isn't really small at all! Here it is mixed with more digestive juices, produced by the intestine and pancreas, and bile, which is made by the liver and stored in the gall bladder. Carbohydrates, sugars, fats, and proteins are all completely broken down in the small intestine.

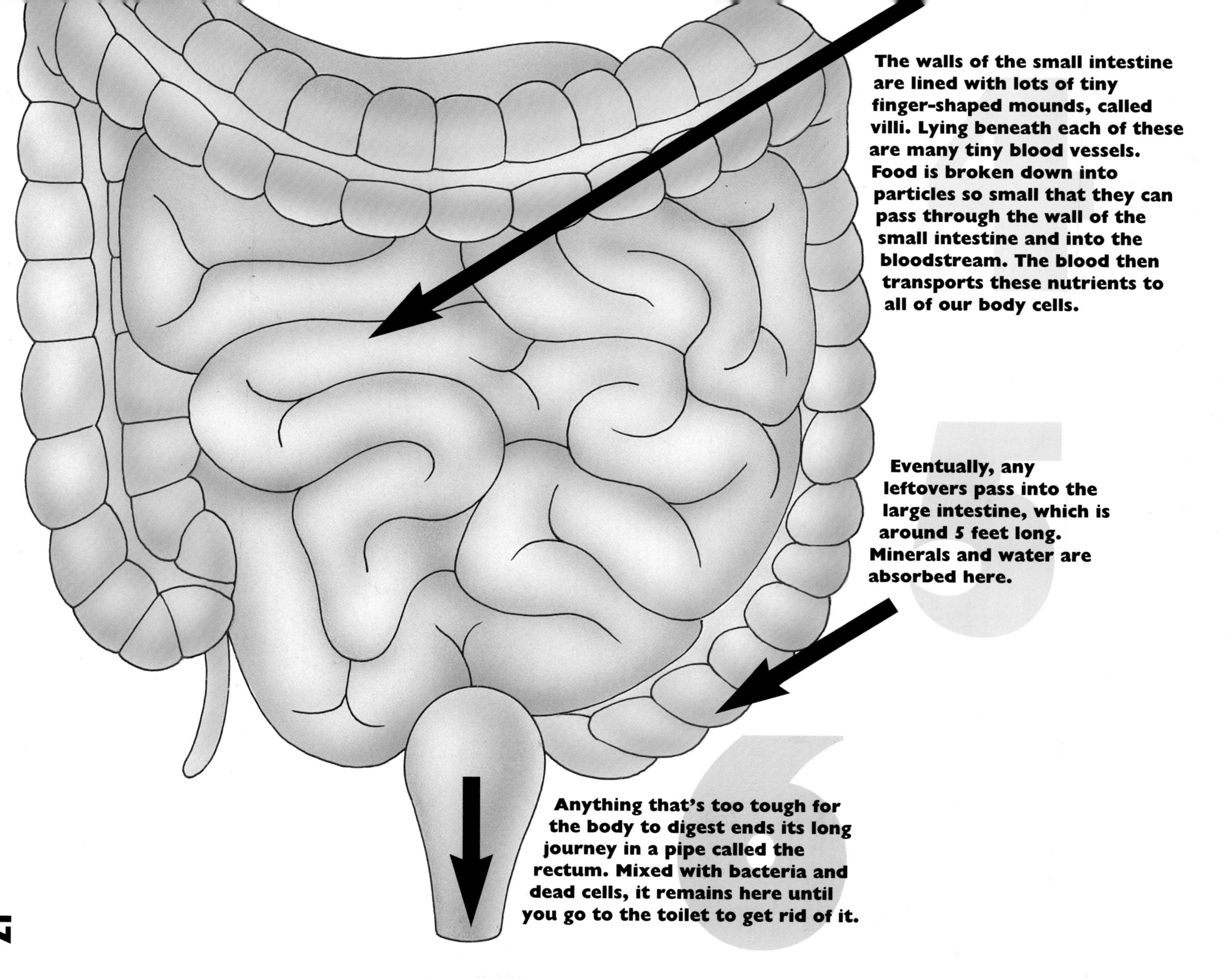

4

The walls of the small intestine are lined with lots of tiny finger-shaped mounds, called villi. Lying beneath each of these are many tiny blood vessels. Food is broken down into particles so small that they can pass through the wall of the small intestine and into the bloodstream. The blood then transports these nutrients to all of our body cells.

5

Eventually, any leftovers pass into the large intestine, which is around 5 feet long. Minerals and water are absorbed here.

6

Anything that's too tough for the body to digest ends its long journey in a pipe called the rectum. Mixed with bacteria and dead cells, it remains here until you go to the toilet to get rid of it.

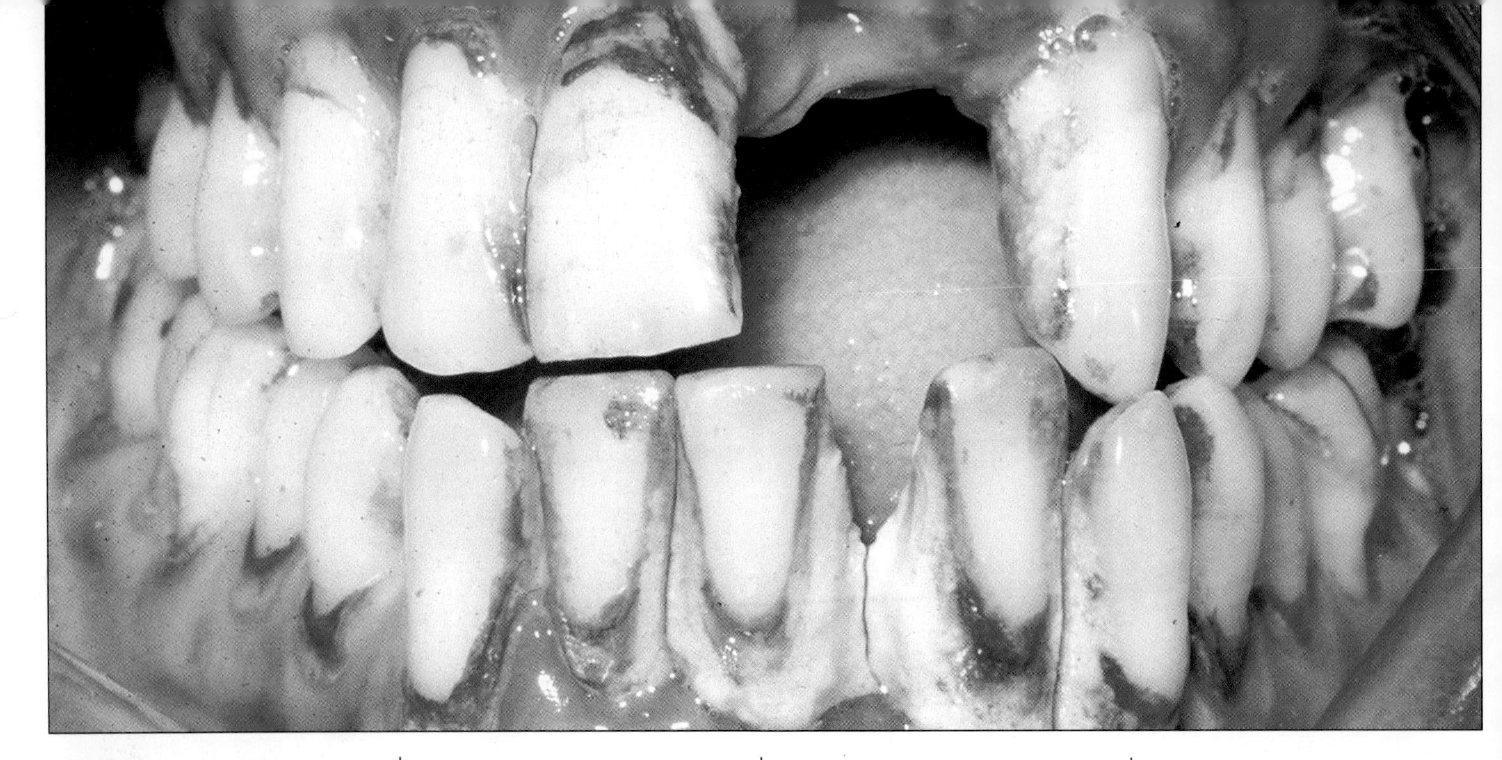

The whole TOOTH

Your teeth may look dead, but in fact they're very much alive. And they can prove it by giving you nasty twinges if you don't look after them. The gruesome teeth in the picture above show what can happen to your pearly whites if you don't lavish tender loving care on them!

You may already have lost 20 of your teeth, but don't panic! These baby teeth fall out naturally from around the age of six, to make way for your permanent teeth.

If you're lucky you'll have 32 of these by the time you're in your twenties, including 4 wisdom teeth. But if you don't develop wisdom teeth, it doesn't mean you are any less clever than your friends – today's humans don't really need them.

TAKE A BITE

Our ancestors had huge jaws and used their gnashers to chomp on raw meat and hard bones. These days you rely on 8 front incisors, with their sharp cutting edge, and four pointed canines to bite off chunks of food.

Behind these, your broad lumpy premolars and molars are used to grind food into a paste you can swallow. All these hard-working teeth are firmly rooted in your jaw and coated in white enamel – the hardest substance in the human body.

BEASTLY TEETH

Mammals' teeth tend to differ according to their usual dinner.

- Plant-eating mammals such as sheep and giraffes have big, flat molars that are perfect for chewing leaves and stalks.
- Meat-eating mammals, like tigers and wolves, have long pointed canines, which they use to tear flesh.
- As humans can eat plants and meat, we have a mixture of both types of teeth.
- Sharks have rows and rows of teeth, which move forward to replace any that are lost. As they have up to 12,000 gnashers, it's a good thing they don't have to use toothbrushes!

INSIDE INFO

Hopefully, you won't ever see the insides of your own teeth! So here's a cutaway diagram for you to chew on instead!

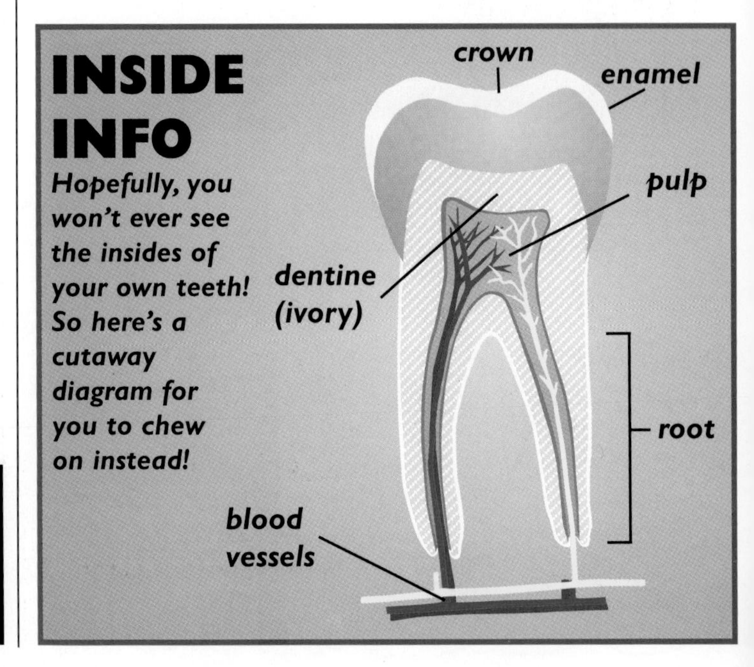

AAAGHH!

Meet the bugs that live in – and on – your body!

You may not be able to see them, but your body is literally crawling with microscopic creatures. In fact, you're home to an entire mini-zoo!

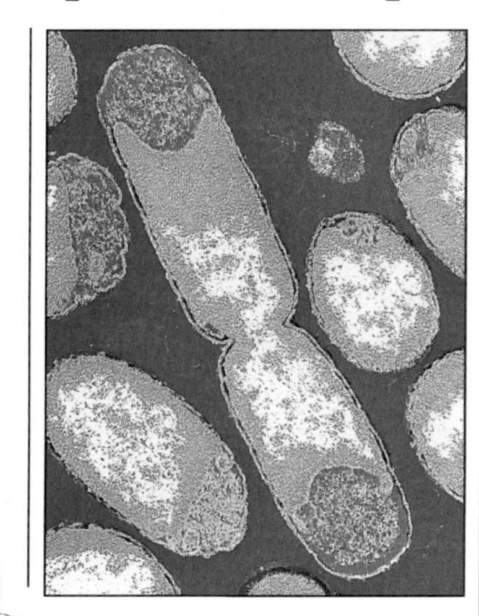

▲ *Microscopic "good" stomach bacteria*

If you didn't clean your pillow for ten years, half of its weight would be made up of dust mites and their droppings!

▲ DEVILS!

Do your eyes get red and itchy whenever you shake out your bedclothes or sweep the floor? You're probably a victim of the dastardly dust mite! These minuscule monsters are invisible to the human eye, but that doesn't stop them causing havoc. Even though their daily diet cleans up flakes of dead skin and sweat, their droppings cause nasty allergies.

◀ BACTERIA ALERT!

The inside and outside of your body are covered in lots of almost invisible bacteria. Most of these mini-creatures do you no harm and some are pretty useful. For instance, bacteria in your intestines can help digest food. But it's important to keep this "good" bacteria in balance and avoid letting "bad" bacteria (i.e., germs) that might cause illness invade your body.

▶ BED BUGS

Chomp! Bed bugs are tiny insects that set up camp in bedding, suck blood, and nibble at unsuspecting humans. Gulp!

▼ HAIR BUGS

Hair lice love nice clean tresses where they can happily guzzle blood from the scalp and lay eggs on hairs. Lice often spread quickly from head to head in school classrooms where children work very closely together. Luckily, they're easy to destroy with special shampoo.

EYELASH HORRORS!

Around half the population has a colony of harmless, spidery, minuscule mites lurking in their eyelash follicles. The 0.01-inch bugs cling onto the eyelashes with four pairs of short legs and feed on the natural oil (sebum) that lubricates your skin and hair.

VITAL STATISTICS

The heart, lungs, kidneys, and liver – vital organs with key roles to play in keeping your body fit

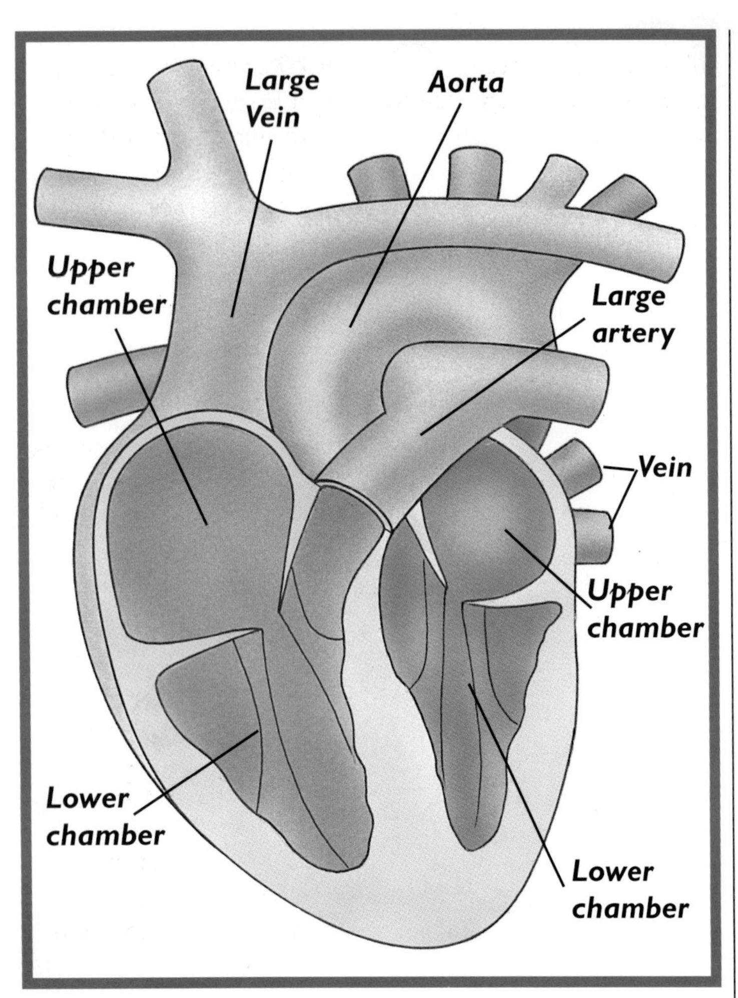

THE HEART OF THE MATTER

Your heart works hard every single second of your life, beating around 100,000 times every day. With each heartbeat, life-giving blood is pumped around the body, fueling your other vital organs.

Every day the heart pumps around 90,000 pints of blood – that's enough to fill over 150 baths!

HOW IT WORKS

The heart is a fist-sized bag of muscle found between your lungs. A dividing wall separates the left and right side into two halves, and each half has an upper and lower chamber.

Four valves in the heart make sure that the blood can only travel in one direction – from the upper to the lower chambers. As the valves slam shut they make a thump-thump sound – that's your heart beating.

Blood from all parts of your body (apart from your lungs) enters the heart via a large vein. It passes through the upper chamber and down to the lower chamber on the right side; then a large artery carries the blood to your lungs where it is refilled with oxygen.

Next, a vein carries the blood back to the left side of your heart where it is pumped from the upper chamber to the lower chamber, then out of the heart and into all of the body's arteries. Thin capillaries return blood to the veins where it is transported back to the right-hand side of the heart. And so on. At the end of 24 hours every blood cell in your body will have visited the heart over a thousand times!

There are about 7 million air sacs in your lungs. If you opened them all out and laid them flat, they could cover a tennis court!

BREATH OF LIFE

Your lungs are vital as they bring oxygen, the gas needed by all human cells to produce energy, into your body.

HOW THEY WORK

When you breathe in through your nose or mouth, your chest muscles pull your ribs up and out and the diaphragm muscle flattens. As your chest expands, air rushes down your windpipe into two branching tubes that travel into your lungs. These divide into smaller airways that are like the twigs of a tree and end in air sacs covered in

You can only talk when you are breathing out – try breathing in and talking at the same time!

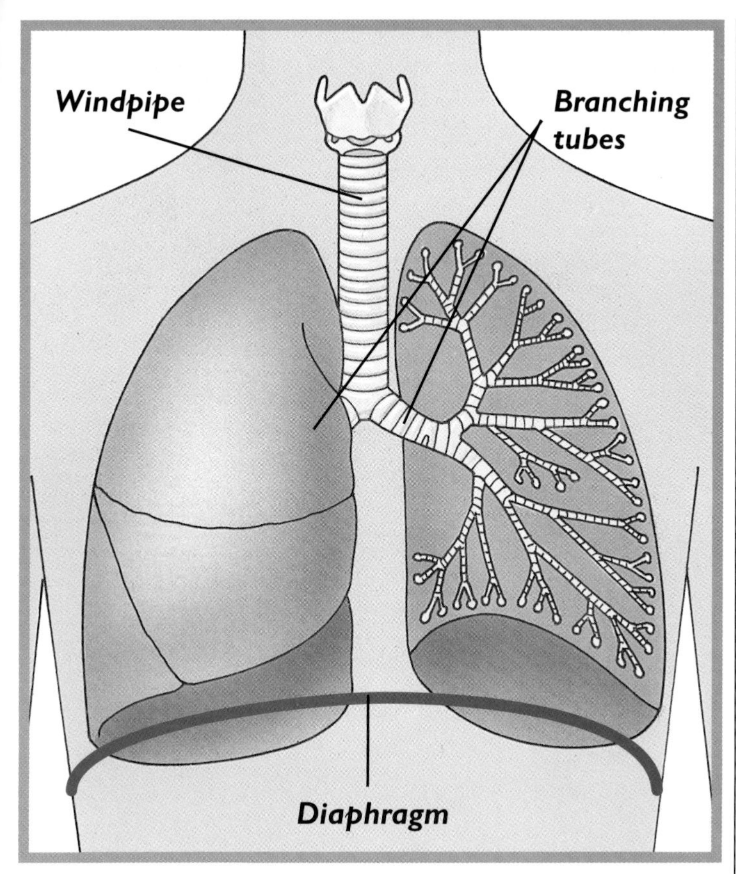

tiny blood vessels. Oxygen passes through the thin walls of the sacs into the blood and is carried to the body cells.

In return, waste carbon dioxide passes from your blood into the sacs. Then when your ribs return to their normal position and the diaphragm muscle rises, the lungs are squeezed, and you breathe this gas out.

The left kidney is higher than the right one because the liver pushes the right one down.

CLEANING MACHINE

Your two bean-shaped kidneys can be found in the middle of your back and are responsible for cleaning all the blood in your body. Blood passes through each kidney 400 times every day. Kidneys remove any waste products that build up after chemical reactions have taken place in your cells.

HOW THEY WORK

Every minute, over 2 pints of blood flow through arteries into your kidneys. Here, the arteries divide into smaller channels through which the blood flows much more slowly.

Each of these smaller channels ends in a blood filtration unit called a nephron. There are about two million nephrons in the kidneys and they filter around 3,000 pints of blood a day. As blood flows though them, substances needed by the body (such as sugar and minerals) are absorbed back into the blood through special tubules, while waste is collected and transported to your bladder.

BUSY BUSY BUSY!

Your liver is your body's largest and busiest internal organ. Not only does it help your body digest food, but it also stores it and filters poisons and waste from the blood.

Your liver could still function if three-quarters of it was removed!

HOW IT WORKS

The liver consists of two sections called lobes, and each lobe is made up of around 50,000 storing and absorbing units.

The liver receives blood containing oxygen from the heart and also blood filled with nutrients from the small intestine. In the liver the main artery and vein branch into a network of tiny blood vessels. At the end of these, spongy cells absorb the nutrients and oxygen, and waste is filtered out. Meanwhile vitamins, minerals, and other goodies are added to the blood and carried back to the heart.

Your liver also makes about 2 pints of bile every day. This is a green substance that helps to break up and digest fats in the small intestine. Some trickles straight into your small intestine, and the rest is stored in your gall bladder to be released when food arrives.

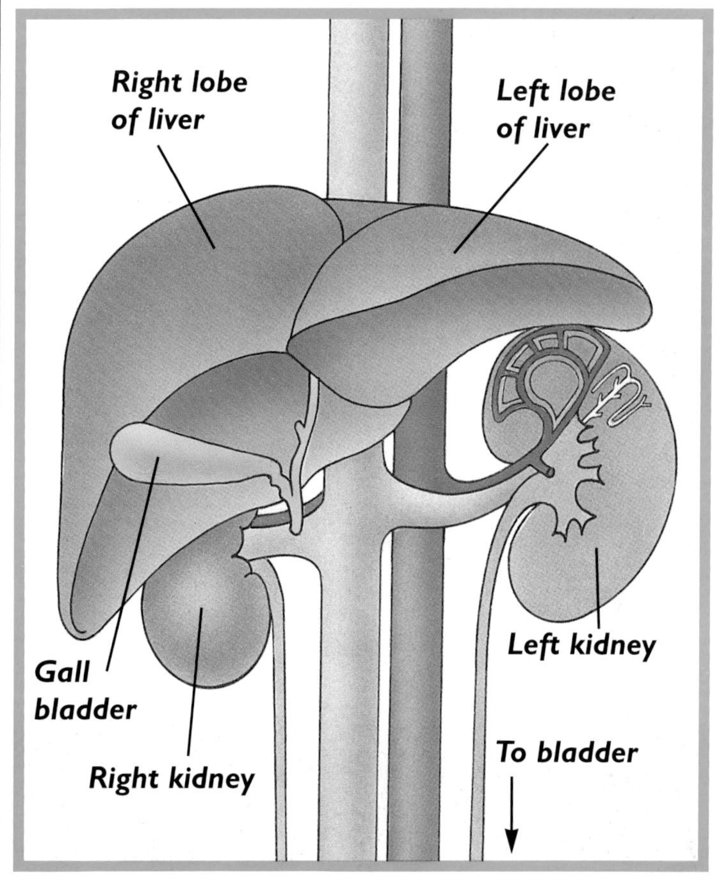

What on earth's going on?

They may feel strange and sound even stranger, but reflexes, like sneezing, coughing, hiccuping, and yawning are your body's way of sorting itself out!

SNEEZING FIT!

You get that urge to sneeze when something is irritating your nose. Thousands of tiny hairs line the walls of your airways, and these keep dust particles out of the nose. But if things become clogged, sneezing gives your airways a quick, thorough clean.

First you breath in deeply, then air passes violently through your mouth and nose, blasting out the offending particles. Get your hankies out!

Like sneezing, coughing is a way of cleaning out your airways. This automatic reaction helps clear irritating mucus or particles out of your lungs.

HIC! COUGH! A-CHO

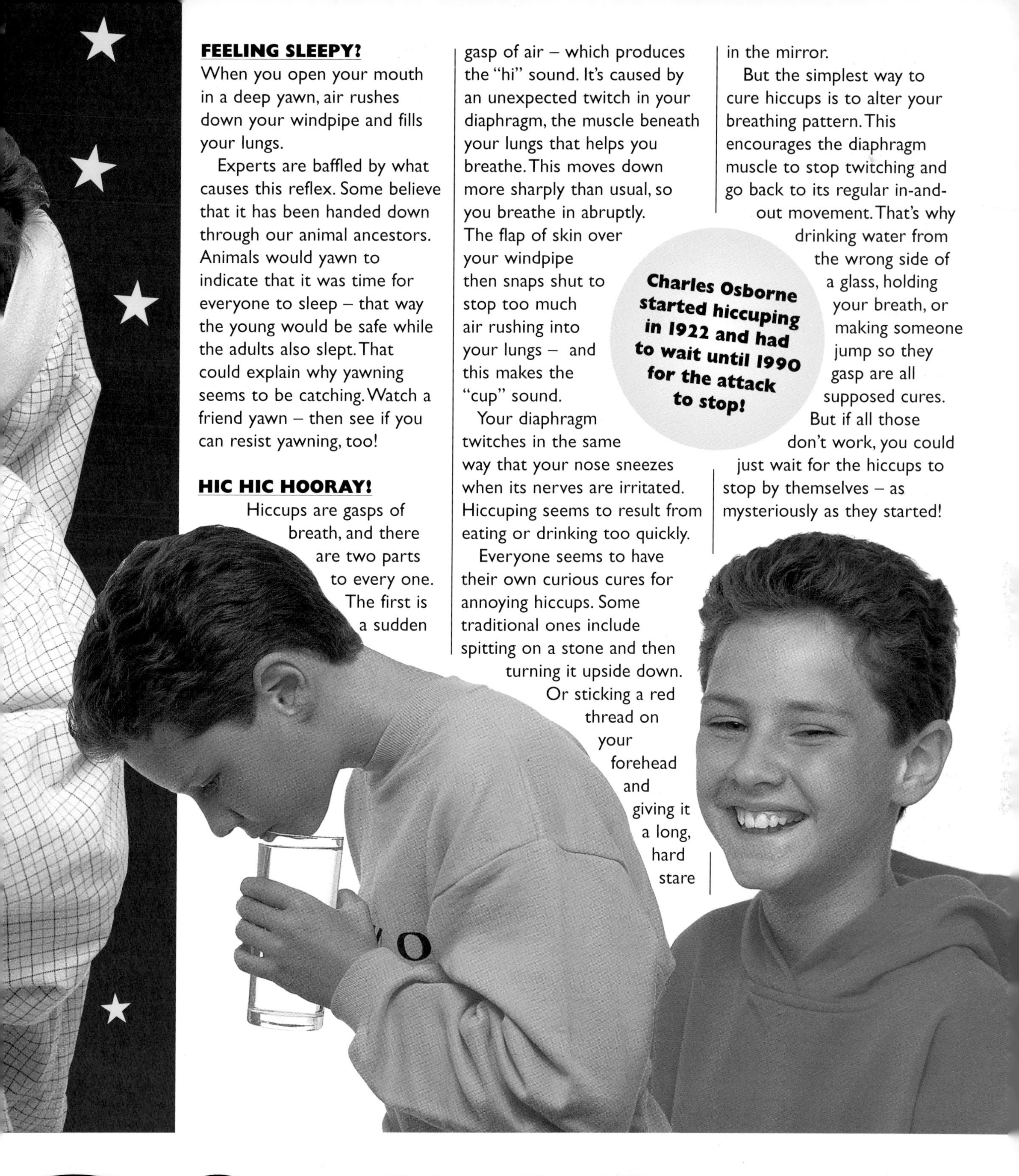

FEELING SLEEPY?

When you open your mouth in a deep yawn, air rushes down your windpipe and fills your lungs.

Experts are baffled by what causes this reflex. Some believe that it has been handed down through our animal ancestors. Animals would yawn to indicate that it was time for everyone to sleep – that way the young would be safe while the adults also slept. That could explain why yawning seems to be catching. Watch a friend yawn – then see if you can resist yawning, too!

HIC HIC HOORAY!

Hiccups are gasps of breath, and there are two parts to every one. The first is a sudden gasp of air – which produces the "hi" sound. It's caused by an unexpected twitch in your diaphragm, the muscle beneath your lungs that helps you breathe. This moves down more sharply than usual, so you breathe in abruptly. The flap of skin over your windpipe then snaps shut to stop too much air rushing into your lungs – and this makes the "cup" sound.

Your diaphragm twitches in the same way that your nose sneezes when its nerves are irritated. Hiccuping seems to result from eating or drinking too quickly.

Everyone seems to have their own curious cures for annoying hiccups. Some traditional ones include spitting on a stone and then turning it upside down. Or sticking a red thread on your forehead and giving it a long, hard stare in the mirror.

But the simplest way to cure hiccups is to alter your breathing pattern. This encourages the diaphragm muscle to stop twitching and go back to its regular in-and-out movement. That's why drinking water from the wrong side of a glass, holding your breath, or making someone jump so they gasp are all supposed cures. But if all those don't work, you could just wait for the hiccups to stop by themselves – as mysteriously as they started!

Charles Osborne started hiccuping in 1922 and had to wait until 1990 for the attack to stop!

O? YAWWWN

TEARS AND LAUGHTER

A sob story with a happy ending!

Sniffle, sniffle, sob, sob. Can you remember the last time you had a really good cry? It may seem babyish to bawl, and many people believe in holding back their tears, but they're probably doing themselves more harm than good!

WAIL I NEVER!

Tears really do make you feel better. Your eyes are making tears to clean themselves all the time, but scientists have found that the tears you produce when you are miserable are different from the ones you produce when your eyes are just irritated.

Your emotional tears contain a substance that makes you feel depressed, and when you cry you get rid of it and feel better.

It can actually be unhealthy to hold back your tears. If you don't express your feelings and get things out of your system, you risk becoming more upset – and it can even have a bad effect on your body's immune system. You may become more vulnerable to nasty bugs, and you're more likely to get ill. So don't store up your feelings when you're upset – find a quiet spot and have a good cry!

Humans are probably the only animals that cry when they're upset or in pain – apart from a soppy sea mammal called the dugong.

WHAT ARE TEARS MADE OF?

Tears are made by special glands located behind the eyelids. Your eyes blink around 20,000 times a day, and each time they coat the outer layer of your eyeball with fluid. This liquid is mostly a salt solution but it also contains substances that fight bacteria and keep your eyes clean. Tears also protect your eyes from strong fumes – which is why they water when you're chopping onions.

Tears flow out of your eyes through ducts at the inner corner of each eye. They drain into a larger tube before emptying into your nose. If you're having a big cry, tears run out of your nostrils, too!

Babies don't have their first giggle until they're around four months old!

A LAUGHING MATTER

Laughter is a great "let-go" – you chuckle when you find something funny, and your body releases "feel-good" chemicals that reduce stress. And if you can see the funny side of things – whether it's a practical joke your little brother plays on you or tripping over your shoelaces in class – it can stop you from getting uptight.

A regular giggle can be good exercise, too. A big belly laugh exercises your main nerves, organs, and muscles. In fact, it's possible for around 400 of your muscles to move when you laugh. That means 100 belly laughs would be equal to ten minutes' cycling, rowing, or jogging. What a workout! But it seems children would beat adults hands down in a laughter competition. The average child giggles around 150 times a day, whereas the average grown-up laughs just six times a day. Tee hee!

HAIR The long and the short of it

RAISING FACTS!

● Everyone is unique, but you have roughly 100,000 hairs on your head that protect your scalp from the sun.

● Most of the human body is covered in tiny, soft hairs, and only certain areas, such as the palms of the hands, have absolutely no hair at all.

● The hair you can actually see consists of dead cells – that's why it doesn't hurt when you cut it.

● Each hair starts off alive in a deep pit called a follicle. As it grows, it moves upward as new cells form beneath it. The hair is eventually cut off from the blood supply, dies, and turns into a hard protein called keratin in the process.

● An oil gland in the follicle keeps dead hair soft and supple, but if it produces too much oil your hair looks greasy.

● Scalp hairs grow around 0.013 inches a day, although they tend to grow fastest in the summer months.

● After two to six years of growing, a hair falls out. Then the follicle the hair grew out of rests for about three months before producing a new one.

● Every day we lose around 80-100 hairs.

● If left uncut, hair will usually grow to a maximum of 24–36 inches. But that's nothing compared to the record-breaking locks of an Indian woman called Mata Jagdamba. When her hair was measured in 1994 it reached an amazing 13.87 feet!

● Whether you have curly or straight hair depends on the shape of your hair follicles. Straight hair grows from round follicles, wavy hair from oval-shaped follicles, and springy, curly hair from flat follicles.

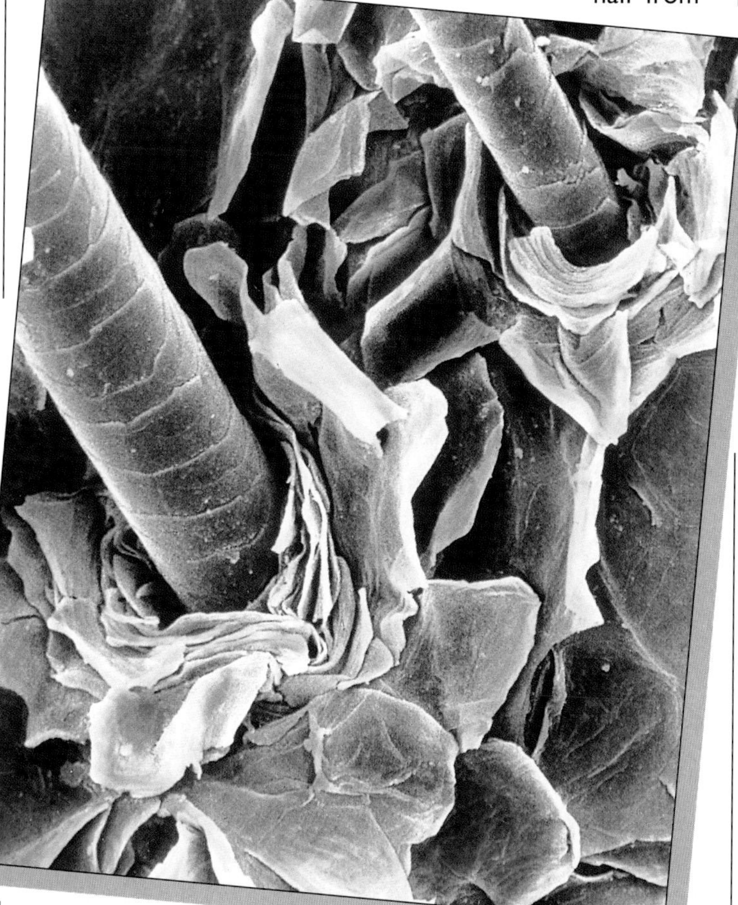

▲ ***Hair we go! Hair follicles magnified 1,320 times***

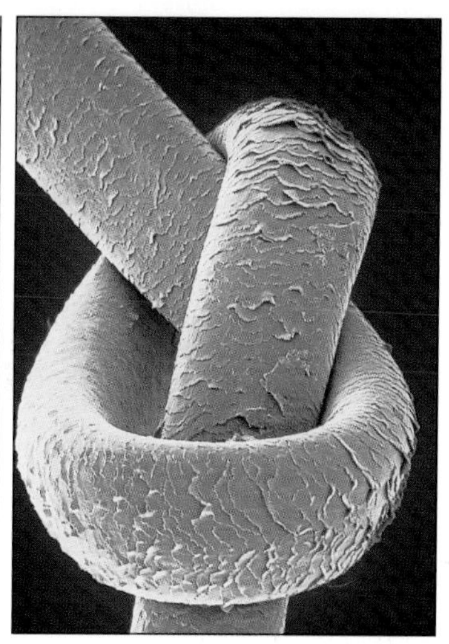

▲ ***Not a knotted ship's rope – a strand of hair magnified 300 times***

● Your hair color is determined by the amount of a substance called melanin that it contains. Dark hair has lots of melanin, whereas blond hair contains very little. As people grow older, their production of melanin slows down, and their hair gradually turns white or gray.

● Blondes have much more hair than redheads.

NAIL'S PACE

Even though they look totally different, your fingernails and hair are made of the same substance – keratin.

SEEING DO

Terrific twins, tremendous triplets, and squads of quads!

▲ *All good things come in threes!*

Can you imagine having a friend for life who's the mirror image of you? They're not always identical, but about one in every ninety babies born in the United States of America is a twin.

Some children are born with a whole gang of pals. Triplets are three babies, quadruplets are four, quintuplets are five, and sextuplets are six — yes, six — babies born together!

BABY BOOM

If the mother's egg splits into two after it's been fertilized by a sperm, a pair of identical babies grow. If she releases two eggs at the same time and these are both fertilized, then two nonidentical (or fraternal) twins develop.

Likewise, if the mother releases three, four, or more eggs then triplets, quadruplets, or even more babies will be born. These could be all girls, all boys, or a mixture of the two. However, identical twins are always both boys or girls.

Twins and other multiple-birth babies usually weigh less than single babies when they're born. But the world's heaviest twins weighed in at a combined 27 lbs, 12 oz in 1924. The bundles of joy were born to Mrs J. P. Haskin of Fort Smith, Arkansas. She certainly had her hands full!

DOUBLE ACT

As well as having similar physical characteristics, many twins, triplets, etc., reckon that there is an extra-special bond

Multiple births are very common in the animal kingdom – most creature moms have more than one baby at a time.

▲ *No, we're not taking the Mickey! This six-pack is a set of sextuplets.*

▶ *These twin sets are all in the same class in elementary school in England. Talk about seeing double!*

UBLE!

DOUBLE TAKE

Everyone does a double take at New York City's Twins Restaurant. No, you don't have to be one half of a twin to eat there, but you're sure to be served by a double act, as all the waiters and waitresses are identical twins!

between them. They say they know what the other is thinking and feeling when they are apart – this is called telepathy.

Some twins even claim they get sympathetic twinges if their "other half" is in pain or upset. Scientists aren't convinced that this telepathy actually happens. They say the twins may just know each other really well because they have very similar bodies and have grown up closely together.

But check out this quirky coincidence. When identical twins Lavinia and Lorraine Christmas decided to surprise each other with a festive visit, their cars crashed into each other! The pair ended up in the same hospital room, but luckily they weren't seriously hurt.

TOO CLOSE FOR COMFORT

Some identical twins are born joined together. When their mother's egg splits, it doesn't separate completely, so they end up being linked together and even share some parts of their bodies. These babies are called Siamese twins, named after the famous Bunker babies from Siam (now Thailand).

Chang and Eng Bunker were born joined at the chest in 1811. But even though they faced a lifetime of incredible obstacles, they managed remarkably. They both lived long, happy lives, got married, and even had children.

Modern medical techniques mean that these days Siamese twins can often be separated.

GETTING BETTER
All the time
Miracles of modern medicine

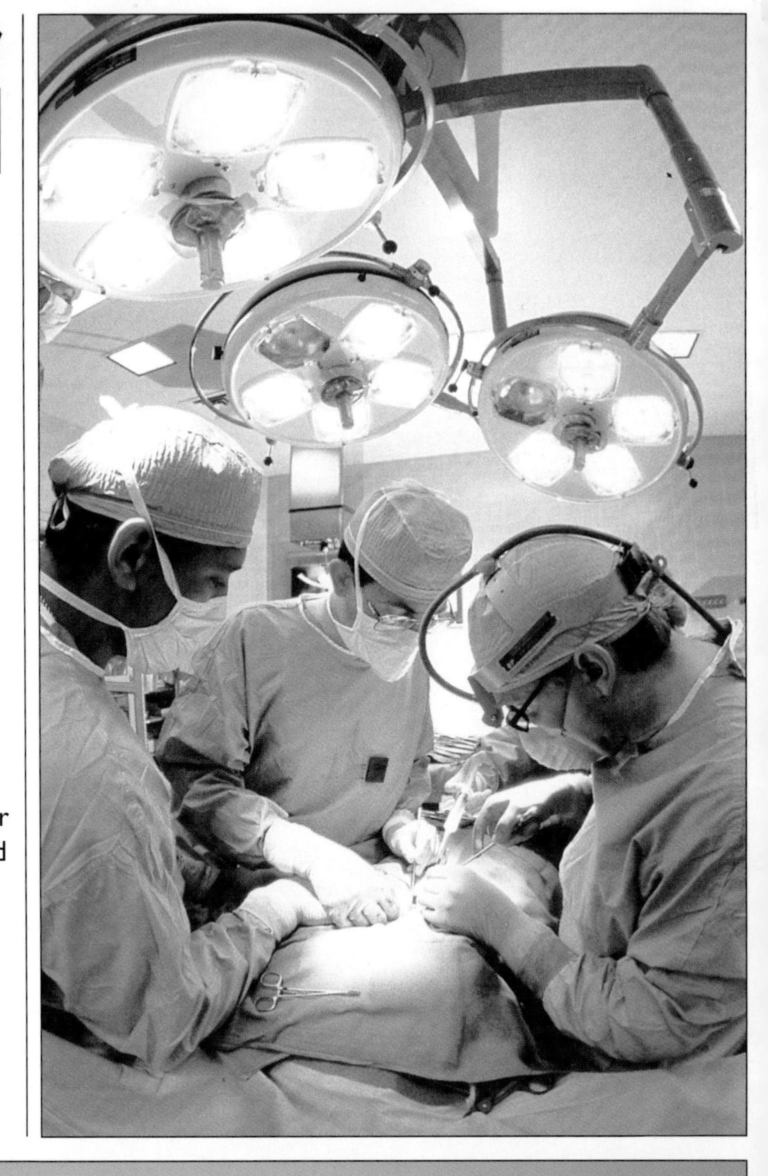

▶ **HEART TO HEART**

In 1967, a South African surgeon named Christiaan Neethling Barnard performed the world's first human heart transplant. With a medical team of 30, he removed the heart from a 25-year-old woman who had died in a car accident. This was placed in the chest of a 55-year-old man whose own heart was damaged. Unfortunately, the patient died of a lung infection 18 days later, but many of today's patients who are given heart transplants live for many years.

NEW FOR OLD

Since the first kidney transplant was carried out in the United States in 1950, doctors have perfected their transplant techniques and developed drugs that stop patients from rejecting organs. Now, they can take liver, eye corneas, and even skin from dead donors and give them to patients with damaged organs. Skin is often grafted onto people who've suffered serious burns.

Several heart-lung-and-liver transplants have been carried out. And in 1994 a 32-year-old British man survived a mind-boggling operation to replace six of his organs at one time – he was given a new liver, kidney, stomach, duodenum, small intestine, and pancreas!

LOOK – NO NEEDLES!

If the sound of the dentist's drill is enough to make you sweat, the latest dental breakthrough will be a huge relief. The KCP2000 is a microblaster that shoots out millions of tiny particles of aluminum oxide to break down tooth decay and spray it away. And best of all, because the KCP2000 doesn't cause any heat or vibration, you don't need an anesthetic. So no injections – hurrah!

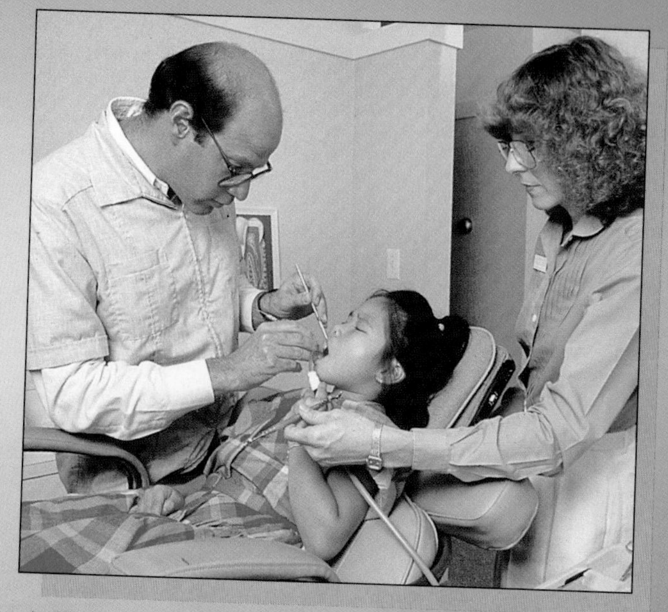

▲ ***This won't hurt a bit – honestly!***

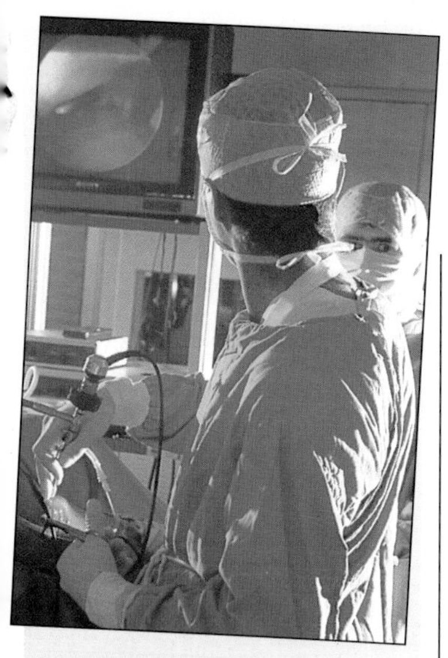

▲ THROUGH THE KEYHOLE

In many of today's operations, surgeons no longer cut a big hole in their patients, so scars may be marks of the past. Instead, surgeons insert a miniature camera through a small incision (cut). Long-handled instruments are then inserted through other small holes, and the surgeon watches what is happening inside the body on a TV monitor. Known as keyhole surgery, these techniques mean patients have to spend less time in the hospital recovering after an operation.

▼ LIGHT FANTASTIC

Lasers are one of the wonders of modern medicine. Lots of short-sighted people have been able to throw away their glasses because lasers are being used to correct their eyesight – permanently. Surgeons are also using lasers to shatter kidney stones, but the latest lasers look set to cut through bones as well. So that may mean an end to surgical drills and saws. It certainly sounds a lot less painful!

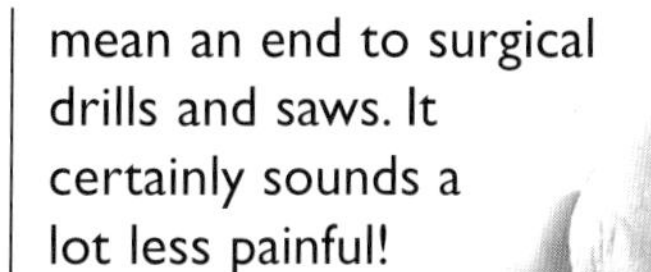

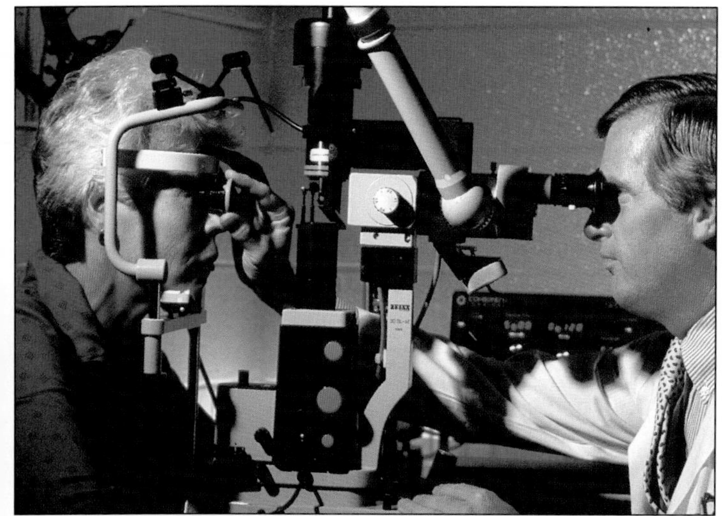

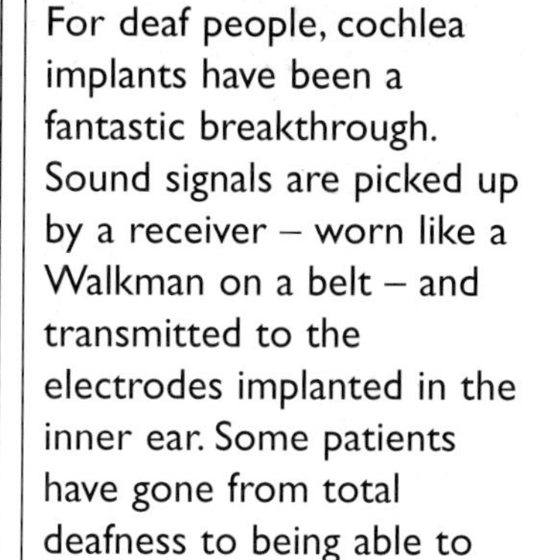

◄ NOW HEAR THIS!

For deaf people, cochlea implants have been a fantastic breakthrough. Sound signals are picked up by a receiver – worn like a Walkman on a belt – and transmitted to the electrodes implanted in the inner ear. Some patients have gone from total deafness to being able to have a conversation on the phone. Amazing!

▼ BACK TO THE FUTURE

Many people are also turning to age-old remedies to help them stay well. Many of these alternative therapies, such as acupuncture, aromatherapy, and reflexology, have been developed over centuries. They often treat a person as a whole, rather than concentrating on curing a particular ailment. These traditional cures are not only supposed to make you healthy, but happy, too!

WOW!

The human body does some amazing things, but these record-holders really excel themselves!

WHAT A FEAT!
Matthew McGrory of Pennsylvania finds it easy to keep his feet firmly on the ground – they're size 23!

STRANGE LEANINGS
An Indian man named Swami Maujgiri Maharaj didn't sit down for over 17 years. He was performing a penance, but allowed himself the luxury of leaning against a plank while he was sleeping.

TRANCE OF A LIFETIME
On July 30, 1988, Portuguese "human statue" Antonio Gomes dos Santos stood motionless for 15 hours, 2 minutes, and 55 seconds. Maybe he was making a stand about something?

OH BROTHER!
A supermom from Shuya, Russia, gave birth to 16 pairs of twins, seven sets of triplets, and four sets of quads – that's 69 children! It's a wonder she didn't run out of names for them all!

AAAAGGGGHHH!
The world record for screaming is 128 decibels – that's louder than a jack-hammer! The piercing voice belongs to an Australian named Simon Robinson. Shhhh!

CHATTERBOX CHAMP
In September 1990, an Englishman named Steve Woodmore spoke 595 words in 56.01 seconds – that's 643.4 words a minute. Whatatonguetwistingtriumph!

FACE VALUE
Norwegian Hans N. Langseth didn't have many close shaves in his life. When he died in 1927, his beard measured 210 inches!

UP TO SCRATCH
You have a challenge on your hands to beat nail-growing champion Shridhar Chillal. He patiently (and carefully!) grew his nails until his thumb nail measured a ridiculous 50 inches long. Apparently, he last cut them in 1952!

GO FOR IT!

Keep your body's machine in tip-top working order

Your body is a sophisticated machine, so it needs to be serviced properly if it's to perform at its very best and repair itself when parts break down.

PUMP IT UP!

Your heart works nonstop every day, but it still likes a workout from time to time. If you exercise hard enough to make yourself a bit breathless, your heart has to beat faster to pump blood around your body and give your muscles the oxygen they need for energy.

But don't start training for the Olympics right away. Why not get off the bus one stop early and walk a bit farther? Or run up stairs instead of always heading for the elevator? Then you can build up to the marathon running!

TIME FOR BED

Your brain is constantly bombarded with information while you're awake, so it needs a rest. Even while you slumber, it's still ticking, together with the other 24-hour working wonders such as your heart, lungs, and liver.

You certainly have a very busy body, so it's not surprising that it needs regular rest to recharge its batteries. In fact, you spend around a third of your life asleep.

As you get older you need to sleep less and less. A new-born baby sleeps 20 hours a day, b[illegible] time you're 10 yo[illegible] probably only nee[illegible] nine hours a night.

New bone is being made and lost all the time. Children replace their skeletons roughly every two years, so get on the trail of calcium-rich food!

VITAL VITAMINS

Your body only needs minute amounts of vitamins and minerals to stay healthy, but they are still very important. Helping to attack infection and keep your body in good condition and encouraging growth, vitamins are your "care and repair fuels."

Some vitamins can be stored by the body, but others, like Vitamin C, need to be absorbed every day. A balanced diet will contain all the vitamins you need (13 in all). So pile on the fruit and veggies and eat yourself healthy!

SHOPPING LIST

There's a saying that goes "You are what you eat." Well, you might not look like your dinner, but if you fuel yourself with healthy foods you'll certainly be full of beans!

- MEAT, FISH, NUTS, EGGS, AND BEANS
They're all bursting with protein, which your body uses to build and repair itself.
- BREAD, CEREAL, AND POTATOES
These are fiber foods that fill you up, give you energy, and get your digestive system into gear.
- FRUIT AND VEGETABLES
More fiber providers and packed with magic minerals and vitamins.
- MILK AND DAIRY PRODUCTS
Packed full of calcium, which is vital for strong healthy bones.
- FATTY AND SUGARY FOODS
Fats are full of energy, but your body only needs a very small amount. The rest is stored as fat, which can leave you very plump! As for sugar, too much can rot your teeth (Check out the picture on page 18 if you want proof!), so keep those cookies and candies for special treats!

INDEX